Nature's Cycles IN REVIEW

THE WATER CYCLE

By Theresa Emminizer

Enslow PUBLISHING

Please visit our website, www.enslow.com. For a free color catalog of all our high-quality books, call toll free 1-800-398-2504 or fax 1-877-980-4454.

Library of Congress Cataloging-in-Publication Data

Names: Emminizer, Theresa, author.
Title: The water cycle / Theresa Emminizer.
Description: Buffalo, New York : Enslow Publishing, 2024. | Series: Nature's cycles in review | Includes bibliographical references and index.
Identifiers: LCCN 2023008316 (print) | LCCN 2023008317 (ebook) | ISBN 9781978536425 (library binding) | ISBN 9781978536418 (paperback) | ISBN 9781978536432 (ebook)
Subjects: LCSH: Hydrologic cycle–Juvenile literature.
Classification: LCC GB848 .E66 2024 (print) | LCC GB848 (ebook) | DDC 551.48–dc23/eng/20230222
LC record available at https://lccn.loc.gov/2023008316
LC ebook record available at https://lccn.loc.gov/2023008317

Published in 2024 by
Enslow Publishing
2544 Clinton Street
Buffalo, NY 14224

Portions of this work were originally authored by Santana Hunt and published as *The Water Cycle*. All new material this edition authored by Theresa Emminizer.

Designer: Joy Broadhead
Editor: Theresa Emminizer

Photo credits: Cover, p. 1 Stanislavskyi/Shutterstock.com; p. 5 Ermolaev Alexander/Shutterstock.com; p. 7 (top) historiasperiodicas/Shutterstock.com; p. 7 (bottom) Lichtfreibeuter/Shutterstock.com; p. 9 24K-Production/Shutterstock.com; p. 10 SUKJAI PHOTO/Shutterstock.com; p. 11 AlexReut/Shutterstock.com; p. 12 BlueRingMedia/Shutterstock.com; p. 13 Soumyo Biswas/Shutterstock.com; p. 14 Roxana Bashyrova/Shutterstock.com; p. 15 Triff/Shutterstock.com; p. 16 Mike_O/Shutterstock.com; p. 17 ND700/Shutterstock.com; p. 19 J.J. Gouin/Shutterstock.com; p. 20 Artiste2d3d/Shutterstock.com; p. 21 VectorMine/Shutterstock.com; p. 23 Wim Hoek/Shutterstock.com; p. 24 DCrane/Shutterstock.com; p. 25 Kertu/Shutterstock.com; p. 27 FloridaStock/Shutterstock.com; p. 28 Stuedal/Shutterstock.com; p. 29 fizkes/Shutterstock.com; p. 30 Artisticco/Shutterstock.com.

Printed in the United States of America

CPSIA compliance information: Batch #CS24ENS: For further information contact Enslow Publishing, at 1-800-398-2504.

CONTENTS

Words in the glossary appear in **bold** the first time they are used in the text.

Water, Water Everywhere!

Water is all around us. It's in the clouds above us. It's in the earth beneath our feet. It's even inside our bodies! Water is always moving between its different forms. It moves in a cycle, or set of steps. This cycle helps us understand these changes!

Learn More

Water makes up about 65 percent of your body! Kids have more water in their bodies than adults do. Adults' bodies are made up of about 55 to 60 percent water.

Different States

Water moves around Earth in different states, or forms: solid, liquid, and gas. Solid water is ice. Liquid water is the water that's in oceans, lakes, and streams. Water in gas form is called water vapor. It's in Earth's **atmosphere**.

Learn More

Water changes from one form to another as it moves along the water cycle. Earth's water is always being recycled, or used over and over again in new ways.

120
80
40

THE BLUE PLANET

Earth is sometimes called "The Blue Planet." That's because about 71 percent of Earth's surface is covered in water! Earth's oceans hold about 96.5 percent of the planet's water. This is salt water. The other less than 3 percent of Earth's water is fresh water.

Learn More

More than two-thirds of Earth's fresh water is in ice caps and glaciers! A very small part of Earth's fresh water is found in lakes, rivers, and other bodies of water.

EVAPORATION

Water changes from its liquid to gas form through a **process** called evaporation. During evaporation, heat from the sun warms the liquid water in oceans, lakes, rivers, and other water bodies, causing some of the water to turn into water vapor.

Learn More

***Heat energy** drives evaporation. It causes the **molecules** in water to speed up. Once they're moving fast enough, bonds connecting the water molecules break. The liquid water then becomes vapor.*

TRANSPIRATION

Water also changes from liquid to gas during transpiration. Transpiration happens when the water in the leaves of plants and trees turns into water vapor. About 10 percent of all water vapor in the atmosphere comes from transpiration.

Learn More

Stomata are tiny pores, or holes, on the outer layer of a plant. Transpiration happens when water moves from the soil through a plant and out its stomata, where it evaporates.

CONDENSATION

Warm water vapor rises into Earth's atmosphere. As water vapor rises, the cool air in the atmosphere causes it to change back into liquid form. This process is called condensation. Condensation is one way clouds are made!

Learn More

Clouds form when water molecules mix with dust and other particles (tiny pieces of things) in the air. Fog is another kind of condensation we can see with our eyes.

PRECIPITATION

When a cloud becomes too full of liquid water, the water falls as precipitation. Precipitation most often happens as rain, but it can also be snow, sleet (partly or fully **frozen** rain) or hail (pieces of ice). The precipitation may then collect in oceans, lakes, and streams.

Learn More

At some point, the precipitation that collects in oceans will once again evaporate and become water vapor. All the steps of the water cycle are happening all the time!

Back into the Cycle

Some precipitation just evaporates again and goes back to the atmosphere. In sublimation, snow and ice become water vapor without changing to a liquid first. However, a lot of precipitation flows over land into bodies of water, including oceans, rivers, and lakes. This is called runoff.

Learn More

Fertilizer is something that's added to the soil to make plants grow. Farmers often use fertilizer to grow their crops. When water flows over fields that have been fertilized, or over trash, pollution spreads.

INFILTRATION

When water from precipitation sinks into the soil it's called infiltration. Infiltration can help refill aquifers. Aquifers are underground rock or sand bodies. They hold water that people often use. Some aquifers are very deep and store water for thousands of years!

Learn More

Infiltrated water may also flow up and become part of a body of water. It may be absorbed, or taken in, by plants' roots. It can then return to the cycle through transpiration.

INFILTRATION

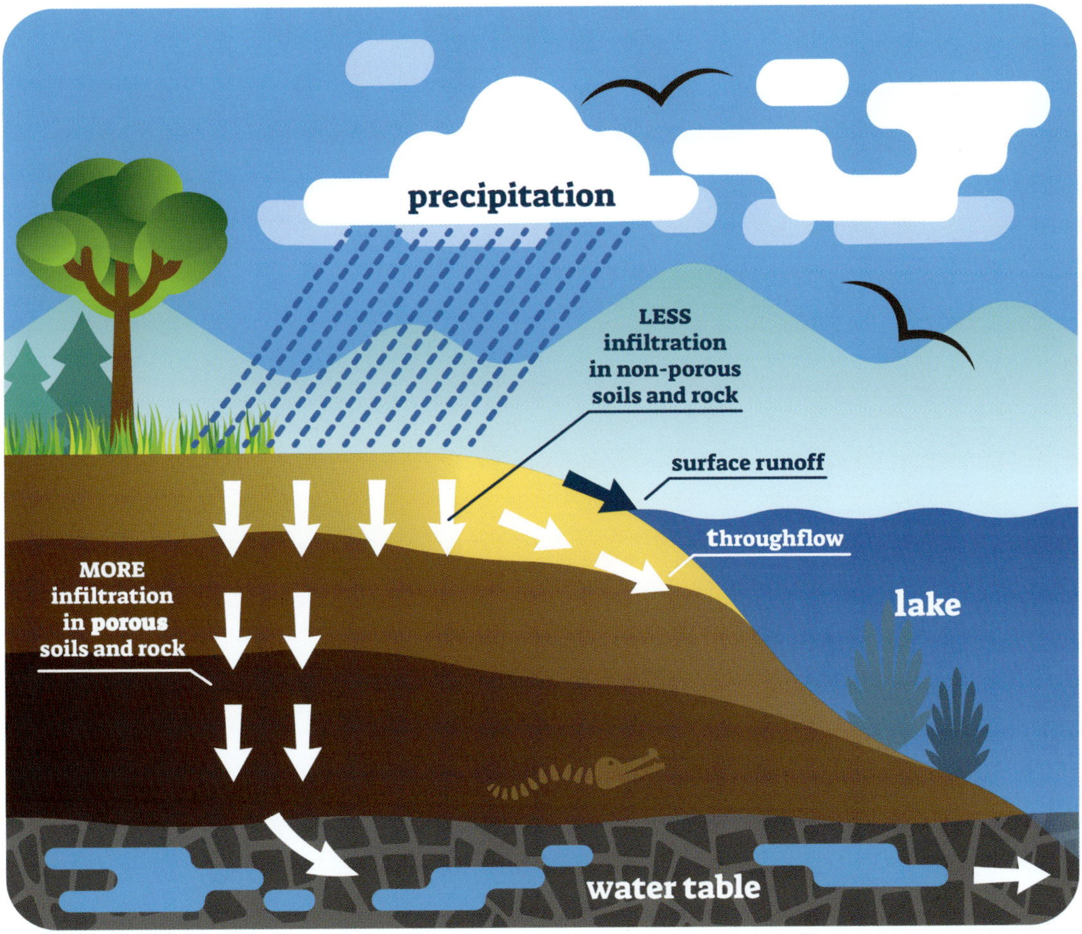

Solid Water

Much of Earth's water is ice. About 90 percent is in Antarctica and 10 percent is in Greenland. Ice is also found in glaciers. Water stored as ice and glaciers doesn't move through the water cycle as surface water does.

Learn More

Glaciers form when more snow falls than melts in an area. About 10 percent of Earth's land is covered with glaciers, but this amount is shrinking. Glaciers move very slowly, shaping the land as they go.

Pools

Water stored in ice and glaciers may stay frozen for thousands of years. It's part of the water cycle, but unlike water in oceans or lakes, it moves through the cycle very slowly. Places where water is locked up or stored are sometimes called pools.

Learn More

Did you know that ice affects the weather? Ice is bright white. It reflects, or bounces back, sunlight. This keeps Earth from getting too hot. However, some human activities are causing ice caps and glaciers to melt.

CHANGING THE CYCLE

Warming weather affects every part of the water cycle. As Earth's **climate** changes by becoming warmer, ice is melting faster, **releasing** more water into oceans. Warmer **temperatures** also make evaporation happen faster, which leads to heavier precipitation.

Learn More

Ocean currents, or flows, move warmer water to cooler areas and cooler water to warmer areas. Warming temperatures have affected these currents, which creates changes to the water cycle and weather worldwide.

WONDERFUL WATER!

Take a sip of water. Think about all the different forms it may have taken before reaching you! Maybe it was in a cloud. Maybe it was underground. One thing's for sure, that water is part of the water cycle!

Learn More

The water cycle is also called the hydrologic cycle. It has been at work moving water from one form to another for over 4 billion years!

STEPS IN THE WATER CYCLE

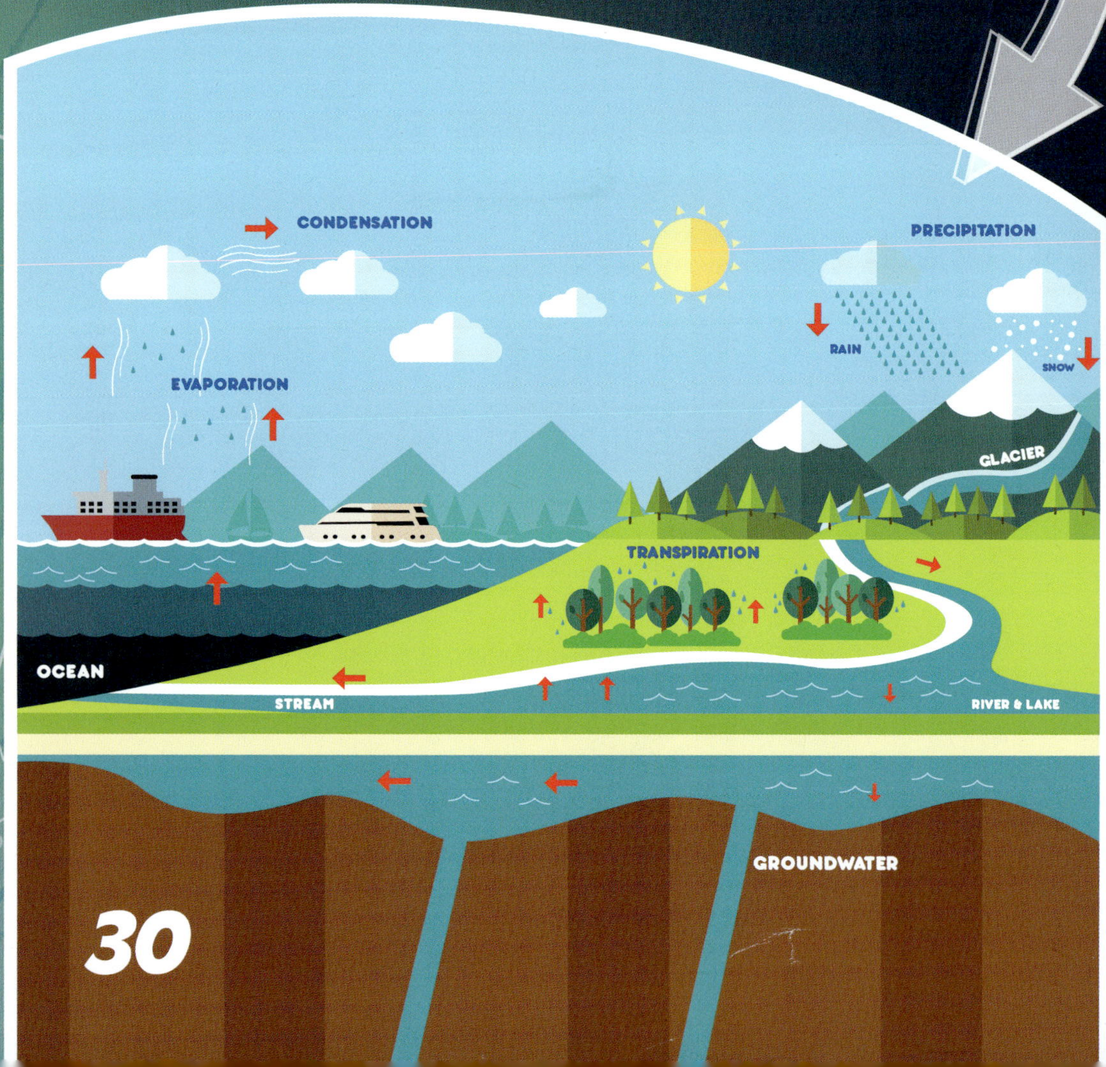

GLOSSARY

atmosphere: The mixture of gases that surrounds a planet.

climate: The average weather conditions of a place over a period of time.

energy: Power used to do work.

frozen: Turned to ice.

glacier: A slowly moving mass or river of ice formed by the gathering and pressing down of snow on mountains or near the poles.

ice cap: A covering of ice over a large area, especially on the polar part of a planet.

molecule: A very small piece of matter.

pollution: Matter that makes water, air, or the land unclean and often no longer safe for people to use.

porous: Having many holes.

process: A series of steps or actions taken to complete something.

release: Let out.

temperature: How hot or cold something is.

FOR MORE INFORMATION

BOOKS

Beasley, Ava. *What's the Water Cycle?* New York, NY: Rosen Publishing Group, 2018.

Heitkamp, Kristina Lyn. *The Water Cycle*. New York, NY: Britannica Educational Publishing, 2018.

WEBSITE

NASA Climate Kids
climatekids.nasa.gov/water-cycle/
Learn more about the water cycle and why it matters!

Publisher's note to educators and parents: Our editors have carefully reviewed this website to ensure it is suitable for students. Many websites change frequently, however, and we cannot guarantee that a site's future contents will continue to meet our high standards of quality and educational value. Be advised that students should be closely supervised whenever they access the internet.

INDEX